Sturgeon Reach

Series Editor: Terry Glavin
Published by New Star Books

Other books in the Transmontanus series

1 A GHOST IN THE WATER *Terry Glavin*

2 CHIWID *Sage Birchwater*

3 THE GREEN SHADOW *Andrew Struthers*

4 HIGH SLACK *Judith Williams*

5 ALL POSSIBLE WORLDS *Justine Brown*

6 RED LAREDO BOOTS *Theresa Kishkan*

7 A VOICE GREAT WITHIN US *Charles Lillard with Terry Glavin*

8 GUILTY OF EVERYTHING *John Armstrong*

9 KOKANEE: THE REDFISH AND THE KOOTENAY BIOREGION *Don Gayton*

10 THE CEDAR SURF *Grant Shilling*

11 DYNAMITE STORIES *Judith Williams*

12 THE OLD RED SHIRT *Yvonne Mearns Klan*

13 MARIA MAHOI OF THE ISLANDS *Jean Barman*

14 BASKING SHARKS *Scott Wallace and Brian Gisborne*

15 CLAM GARDENS *Judith Williams*

16 WRECK BEACH *Carellin Brooks*

17 STRANGER WYCOTT'S PLACE *John Schreiber*

18 OFF THE HIGHWAY *Mette Bach*

19 STRANGER ON A STRANGE ISLAND *Grant Buday*

21 GARDENS AFLAME *Maleea Acker*

Sturgeon Reach

SHIFTING CURRENTS AT THE HEART OF THE FRASER

Terry Glavin
& Ben Parfitt

TRANSMONTANUS | NEW STAR BOOKS VANCOUVER

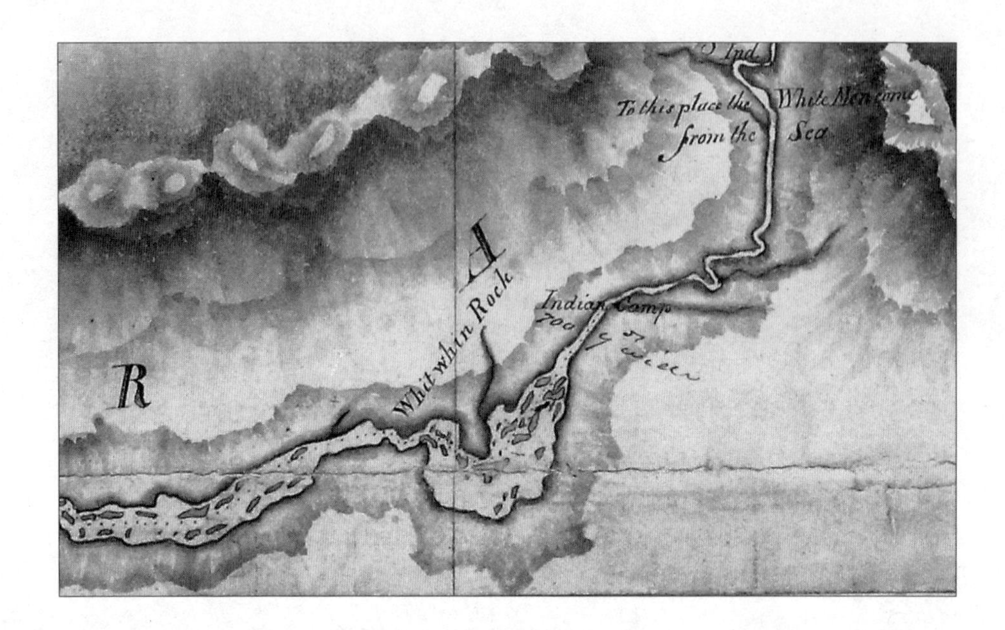

The Fraser River between what is now known as Mission and Yale, early 1800s. Produced circa *1826 by David Thompson and based on maps from his 1814 exploration of the river, this is one of the earliest colonial maps of the Fraser, and the first to give it a name.* COURTESY DEREK HAYES, FROM BRITISH COLUMBIA: A NEW HISTORICAL ATLAS.

CONTENTS

Foreword *by Mark Angelo* 7

Prologue 11

I. A Fully-Loaded Dump Truck Of Gravel For Every Woman, Child And Man 15

II. Spring Bar 20

III. The Stone People 23

IV. Wandering Channels, Giving And Taking, And A Still-Beating Heart 29

V. Memories Of The Great Flood 39

VI. 'Gritting Its Teeth' Rock 47

VII. What Happens At Harrison Knob 49

VIII. Arguments That Don't Hold Water, Arguments That Do 52

IX. The Concrete Facts 59

X. Sinking Like a Stone: Gravel Mining and Reduced Flood Risk in the Heart of the Fraser 65

Notes 68

Acknowledgments 70

A ptarmigan surveys the Fraser Valley from Mount Cheam. STEVE CLEGG PHOTO

FOREWORD

by Mark Angelo

We think we know the Fraser. To many British Columbians, it defines our great province, having shaped the history of this land long before Simon Fraser travelled its course. Its sprawling, rich and magnificent delta is a major reason why so many have settled here. And the river continues to provide many among us with their livelihoods.

Yet most of us know only a small part of its nearly 1,400-kilometre passage, and that's the highly developed, urban stretch of the Fraser near its confluence with the Pacific.

I've long had a special attachment to the Fraser. I first had the opportunity to paddle the river's full length, starting from its headwaters near Mt. Robson, back in 1975. Ever since, I've always looked at the Fraser as the soul of our province. On that initial journey, I realized for the first time that, between the towns of Hope and Mission, just a short distance upstream from the area where most British Columbians live, there is a remarkably special place. It's an area of transition, where the river moves from the canyon to the delta; a part of the Fraser that, in recent years, has become renowned as one of the most productive stretches of river anywhere in the world. *This* is the Heart of the Fraser.

When I first paddled this stretch, I had been on the river for almost a month. And I was struck by the area's natural beauty, especially knowing that I was now so close to Vancouver. As authors Terry Glavin and Ben Parfitt so vividly convey in this book, it is an incredibly unique, rich and delicate place.

This section of the Fraser sustains almost thirty species of fish, including Canada's largest population of white sturgeon, an amazing, dinosaur-like creature that can grow as large as 600 kilograms and live for more than 150 years. As many as 20 million pink salmon spawn in this part of the river, while tens of millions of sockeye travel through it to spawning grounds in upstream tributaries. Rich populations of chum and chinook depend on its backwaters.

From a wildlife perspective, the area is equally impressive. Red-tail hawks, bald eagles, turkey vultures, as well as green and great blue herons, frequent the river's edge. There are Oregon spotted frogs and western red-backed salamanders. Deer, bear, coyotes, beaver, marten, and even cougars can be seen along the shoreline.

Many of Canada's most important cultural sites are also found here. Skowlitz burial mounds, sometimes referred to as the Pyramids of the Fraser, date back more than two thousand years. There are ancient village sites and fortifications along with pictographs and natural formations that embody the original human inhabitants' most deeply held spiritual beliefs.

This amazing stretch of river is also known to some as the Gravel Reach, because of the gravel and cobbles deposited there by the powerful currents emerging from the Fraser Canyon. This book more evocatively dubs it Sturgeon Reach.

More important, however, is the authors' keen understanding of the complex dynamics of the river itself as it wanders across an ever-changing landscape of floodplains, side channels and wetlands, which are screened from those who travel Highway 1 by towering black cottonwoods and cedars. The Heart of the Fraser has been monumentally transformed in the last century. By trying to tame its rambling nature, we have destroyed much of it. Resource extraction, agricultural expansion and land development continue to erode what remains of its rich natural ecosystems. There is still much to save — but we must act quickly.

Sturgeon Reach tells us of this place's beauty and history while also calling on us to do what's required to protect the area's immense natural and cultural values. As of yet, despite the area's global significance, there is no collaborative land-use plan in place. In the absence of such a plan, governments, businesses and individuals too often do what they might, swayed by opportunism and misconception. Unfortunately, this often occurs at the Heart of the Fraser's considerable expense.

To protect this incredibly rich environment, many conservation-minded groups, including the Nature Trust of BC, the Rivers Institute at the BC Institute of Technology, the North Growth Foundation, and Watershed Watch, are working together to increase public awareness about the river's values while promoting the need for an overarching, collaborative land-use plan.

Admirable efforts are also underway by the Nature Trust to acquire and protect key private lands. In addition, there are renewed attempts to bring together First Nations, public and corporate interests in the hope of preserving the natural and historical legacies of this part of the Fraser so that our children and theirs can enjoy the heart of this great river as we have.

Few places on Earth still have such a rich and intact stretch of river so close to a major urban centre. We can ensure this natural treasure remains intact – but only if we act together. No book better conveys the opportunity, and the challenge, than *Sturgeon Reach*.

Mark Angelo is the chair of the Rivers Institute at the BC Institute of Technology. He is an internationally acclaimed river advocate, a recipient of the Order of Canada, and has traveled on and along close to a thousand rivers around the world. He remains a leader in the Heart of the Fraser conservation campaign.

Undated photo of Alfred Cline, Joe Louie and Ed Louie with some Fraser River sturgeon, taken in Sto:lo territory. CHILLIWACK MUSEUM AND ARCHIVES

PROLOGUE

There is no place in the world quite like it. Everything about the place tends to be discussed in superlatives. The last, the best, the biggest, the oldest. The most beautiful place, "a second Eden" to the first Europeans who encountered it on foot, and to the Europeans who approached it, by canoe, a place where monsters flung themselves out of the river and into the air. A strange place.

That any of it still exists at all is a marvel.

We're not talking about some storied, mist-shrouded, inaccessible riparian jungle in the Great Bear Rainforest. The place this story is about lies directly beneath a sprawling urban, suburban and exurban hive of 2.5 million people, in British Columbia's Lower Mainland. It's the place you can see below, out of the corner of your eye, when you're crossing the Mission Bridge. The western edge of it, anyway. Almost all of it lies within the bounds of the floodway dyke system of the eastern Fraser Valley. Most people don't even know it's there.

Sturgeon Reach, you could call it, after those creatures the naturalist John Keast Lord described in his 1866 memoir, this way: "As you are quietly paddling along in a canoe, suddenly one of these monsters flings itself into the air many feet above the surface of the water, falling back again with a splash, as though a huge rock had been pitched in the river by some Titan hand." Fraser River white sturgeon, *Acipenser transmontanus*, the largest freshwater fish in North America, are known to reach weights in excess of 600 kilograms. There are still sturgeon there, too, which is by itself something of a marvel.

The Heart of the Fraser is another term for the place. That will do fine.

Yet another term for the place is the Gravel Reach, which situates the area rather more precisely in both time and in space. It's not the silt-laden estuary, where the Fraser River flows backwards and forwards every day in its own grassy rhythms. It's part of that, but it's above all that. It reaches from, say, the Stave River, near Mission, to an afternoon's clambering above Hope, not far from a camp that archaeologists would call the Milliken site, one of the oldest known sites of human habitation in North America. It's the gravel islands and the gravel-formed maze of side channels and sloughs of the roaring and ever-shifting main channel they outride.

There is a type of poplar tree that takes root in the islands and the floodplains of the place. The trees are called black cottonwoods. They're the biggest poplars on earth. Try to picture a cottonwood tree the height of a twelve-storey apartment building, with a trunk close to twelve metres in circumference and a crown the size of a huge house. Then picture forests of such trees. That is what the Gravel Reach once looked like. It still looks like that, in places. In groves. In remnants.

In 1808, when Simon Fraser passed through this particular section of the river that came to bear his name, he couldn't quite believe what he was seeing. Just the houses of the people were so astonishing, so utterly different, so much bigger than anything he had seen in his arduous five-year journey from Montreal, that he took pains to get their precise measure. Near Matsqui, Fraser came upon a single cedar-planked housing complex twenty metres wide under one roof, six metres above the ground. Fraser paced off its length at 210 metres. That's two football fields long.

The people who built such houses are still here, too, which may be the strangest marvel of all. Despite the smallpox, measles and tuberculosis among a lethal mix of epidemic diseases that carried off almost all of them, the Sto:lo – the people of the river --- are among us yet. And they still draw much of their lifeblood from the Fraser Valley's "great feast bowl," which is what they called the place long before other names came along.

Salmon, more than anything, was what mattered. And there are still great runs of salmon that traverse Sturgeon Reach, passing through the ventricles of the Fraser's beating heart on their homeward journeys. Great shoals of salmon end those journeys within the heart of the river itself. The pink salmon run that spawns within the Gravel Reach is the largest single spawning population of salmon on Canada's west coast. It's easily ten million spawners nowadays. Nobody knows for sure. Some fisheries scientists reckon it could be twenty million.

A canoe on the banks of the Fraser River near Hope, circa 1890. CANADIAN MUSEUM OF CIVILIZATION / COURTESY DEREK HAYES

As it does to the white sturgeon, to spawning salmon the gravel matters. Pink, chum, coho, chinook, steelhead, sockeye – what matters, in the end, is gravel, and cool, clean water. This is as it was in the beginning, at the close of the Pleistocene Epoch. When salmon recolonized this portion of the continent after the long winter of the Ice Age, gravel and water were pretty much all that was here. It was also in those early days, several millennia before Stonehenge or the great pyramids of Egypt, that people arrived. The Milliken site appears to have been occupied, almost certainly by salmon fishermen, as far back as 9,000 years ago.

In the Heart of the Fraser, gravel is underneath everything. This is what Clem Seymour's grandmother wanted him to hear, back in the early 1960s, when Clem was eleven years old. It was during salmon fishing time. Clem was with his grandmother in their small boat, heading upriver to their fishing spot above Seabird Island. She turned off the outboard motor. "I want you to listen," she said. "Listen carefully." At first, there was only the silence of the river, maybe some birds, a distant train. Just silence. Then Clem heard it.

It was a hissing, almost a tinkling sort of sound. It was the sound of gravel, countless little pebbles and stones making their ancient and ongoing downstream journey. The sound was coming up through the hull of the small boat from the bottom of the river. The river was moving, as it was in the beginning, more than 9,000 years before. Always moving, always growing, shifting, transforming everything around it. It was the sound of creation itself. As

Clem would come to know later, when he'd become chief of the Seabird Island Indian band, it was also the sound of destruction.

At the heart of the story contained in these pages, it's the gravel that matters. The rocks, the stones, the boulders and the gravel. It's what lies beneath everything.

I

A Fully-Loaded Dump Truck of Gravel for Every Woman, Child and Man

By the close of the twentieth century, the Lower Mainland's rapid urban growth was adding an average of 140 kilometres of new road every year. To build a kilometre of road through a new suburb, you'll need about 10,000 tonnes of gravel, or about 800 dump trucks' worth of rock. Each new kilometre of road brings with it a host of related and additional "underground services," which is to say sanitary sewer lines, storm drains, and water mains, and for these things you'll need another 8,100 tonnes of gravel, or roughly 624 dump trucks of rock. For each new kilometre of road.

Underground networks of drain tile, sewer and water lines are surrounded by gravel. The lines themselves are made of concrete, the principle ingredient of which is gravel or crushed rock. In the cities and the suburbs, most of the hard surfaces people walk on are made with the stuff. So are the buildings we work in, the roads and bridges we drive on, the airstrips we land on, the rail beds, train tunnels and trestles, and on and on it goes. For each housing unit in a typical three-storey condominium, you'll need somewhere between 46 and 51 tonnes of gravel. For each three-storey wood-frame townhouse, 200 tonnes. For each single-family wood-framed house, roughly 340 tonnes, or twenty-six dump trucks' worth. And so on, up the line.

By the middle of the twentieth century, when urban development started taking off in British Columbia, the demand for quarry-blasted and crushed "aggregate" (used in concrete) and other forms of sand and gravel also took off. Ever since, it's been far and away the highest-volume industrial mineral commodity produced in British Columbia.[1] In the early years of the twenty-first century, aggregate in the amount of roughly 50 million tonnes was being put to use in the province annually, enough to fill BC Place Stadium from the floor to the old domed roof nine times, with gravel left over.[2] This is a fully loaded dump truck for every woman, child and man in the province.[3] About half of this vast amount of material was being put to work in urban growth between Hope and the seaward dykes of Richmond.

All this meant constant pressure to find and mine new sources of rock and sand. But where? By the 1990s the question had come to obsess planners, regulators and gravel companies. A 1996 report for the provincial government's Ministry of Employment and Investment forecast a 10 per cent rise in concrete use in the Lower Mainland over the following four years.[4] The same publication noted that much of the sand and gravel to meet that demand would likely come from the Fraser Valley itself, a region rich in Quaternary sediments laid down over the 10,000-plus years since the great glaciers had beat their last retreat. But what had taken thousands of years to be put in place was becoming less and less available.

The problem wasn't just scarcity. Much of the gravel being shifted around the region moved by truck, and gravel is a very heavy and expensive thing to transport. Poor planning was a problem too. The biggest gravel deposit on the south side of the Fraser Valley lay underneath the Abbotsford airport. Then there was the bothersome dust and noise associated with incoming and outgoing gravel trucks, the traffic disruption, the wear and tear on roads. Nobody really wants to live beside a gravel pit. Most of all, in the typical pattern of natural-resource exploitation, the easiest to access gravel reserves went first. In the Fraser Valley, this meant a future of increasing reliance on aggregates blasted from hard rock, in quarries, which is about a third more expensive than loose gravel mined from pits.

Trucks laden with rock are expensive to move, increasingly so in a world of

Sprawling suburbs, like the Promontory Ridge development (above and facing page), need gravel, and lots of it, putting pressure on the Fraser Valley's gravel deposits. COLIN WELCH PHOTOS

rapidly escalating oil prices. Well before the psychological barrier of $100-per-barrel oil prices was first broken in 2008, the word from BC's gravel industry was that transportation costs were the main limiting factor in gravel production. As the twentieth century gave way to the twenty-first, the industry asserted that the markets of Greater Vancouver and the Fraser Valley could bear gravel truck delivery costs of no more than fifty kilometres, and by barge, no more than 150 kilometres.[5] But as it turned out, such neat calculations weren't cast in stone. With diminished gravel supply and rising global demand, the industry proved capable of moving massive volumes of rock great distances. Like oil companies switching from land to sea and from conventional crude to bitumen, the aggregate industry proved adept at turning to more "unconventional" sources of supply.

Not far from the mouth of the Fraser River, just beyond Burrard Inlet and the busy ship traffic in and out of Vancouver harbour, lies Howe Sound, and just beyond that, the Sunshine Coast. Occupying this relatively short stretch of coastline is one of the largest gravel and sand deposits in North America. By the late 1990s these coastal deposits were the top producer in Canada, at some 3.5 million tonnes every year.[6] Within a decade, the deposit known as the Sechelt Pit, or the Sechelt Mine, was on target to produce 5.5 million tonnes a year.[7] The massive operation was designed with one transportation route in mind — water.

With the deposit located on the land side of the highway that snakes up the misty Sunshine Coast, a tunnel was constructed so that sand and gravel could be conveyed without disrupting traffic to barges and ships anchored in Trail Bay. This neat feat of engineering conveyance was later repeated at deposits farther up the coast. The smaller barges could carry up to 1,800 tonnes of aggregate, the larger ones three times that volume. But it was the waiting ships whose holds took the most rock; the biggest could move 72,000 tonnes, and at a fraction of the cost of trucks.

By the new millennium, gravel shortages far beyond the Lower Mainland spurred an explosion of aggregate development up and down the BC coast, as a glowing 2007 report to would-be investors in Polaris Minerals Corporation explained. At the time, Polaris held a majority interest in the Orca Sand and Gravel Quarry near Port McNeill, on northeast Vancouver Island. The company estimated the gravel deposit at 121 million tonnes, an amount that would take a quarter century to deplete. Polaris also owned a majority interest in the Eagle Rock Quarry near Port Alberni, a Vancouver Island community famed for its long inlet that was once a busy transit route for commercial fishing vessels and freighters loaded with forest products bound for distant markets in Japan and other Pacific Rim countries. In addition, it had applied for a "licence of occupation" for a nearby gravel deposit named Cougar.

Polaris was also majority owner of a port facility in San Francisco, where it stored and later distributed gravel from BC, and the company was eyeing a second terminal facility in California's Redwood City. "The economics of shipping aggregates from Vancouver Island to terminal facilities in California are incredible," the investor report enthused. "Amazingly, the cost to ship from the Orca Quarry to San Francisco Bay, a distance of about 1,200 miles, is the same as trucking the material 25 miles, approximately $5 per ton."[8]

Orca wasn't alone in demonstrating just how cheaply and easily piles of rock could be moved around by water. On Canada's east coast, New Brunswick companies were by the new millennium shipping gravel to markets in South America. Back in BC, ships' crews were soon moving aggregate through Juan de Fuca Strait for ocean journeys half way across the Pacific Ocean, to Hawaii.

All the while, the expansion of suburban developments in the Fraser Valley persisted. With every year's passing, the aggregate industry, municipal planners, and the construction industry looked to the Heart of the Fraser itself as a major source of supply. While the international trade in crushed rock and gravel percolated along, with only passing mention in the pages of the occasional business journal, a much different picture was unfolding in Sturgeon Reach.

II

Spring Bar

In January 2008, a backhoe operator made the first of thousands of scoops into a mid-river gravel bed off Spring Bar, a gravel island in the Fraser River near Agassiz roughly half the size of Vancouver's famed Stanley Park. The operation was just one small part of a much larger five-year plan to remove more than two million tonnes of glacially derived and river-deposited gravel from the Heart of the Fraser itself.

Spring Bar is just downstream from the place Clem Seymour, in his grandmother's boat, listened to a strange hissing sound coming from the bottom of the river, all those years before. When the backhoes moved in, they opened deep fissures among and between governments, local farmers, aboriginal people, environmentalists, and salmon anglers. The lines were drawn in the most unconventional ways — old allies were arrayed against each other, old adversaries found themselves on the same side. The Spring Bar operation was a joint project of Seymour's Seabird Island Indian band and Jake's Contracting of Chilliwack.

In the bitter controversy that erupted at Spring Bar, it became almost impossible to sort through the conflicting answers to all the questions the operation raised. Was this really necessary to prevent massive flooding and erosion? Was it just a money grab? Why was the federal fisheries department approving a project that was carrying away salmon spawning habitat in the backs of dump trucks? Why did the provincial government spend $564,000 to pay for a temporary bridge to a private gravel-mining operation?

Clem Seymour at the Seabird Island Indian Band's offices east of Chilliwack, on the other side of the Fraser River. CHARLES CAMPBELL PHOTO

While the arguments raged, thirteen huge off-road trucks and three excavators worked twenty-four hours a day, six days a week, all winter. They dug up a lot more than just gravel during those weeks. They opened up deeply troubling questions about the implications of development within the most important artery of what is still one of the richest and most productive salmon rivers on earth, the last, best wild place in the Fraser Valley, the epicentre of the west coast's biggest pink salmon spawning grounds, and the home waters of *Acipenser transmontanus*, the monsters of John Keast Lord's journal of 1866.

For Clem Seymour, it was all very interesting, but rather more straightforward than one might imagine from reading the newspaper headlines. The Seabird Island reserve was once well in excess of 4,500 acres, huge in comparison to most Sto:lo reserves. Not any more.

"We've lost 1,200 acres to that river. We're still losing five to ten acres a year. We have lost twenty or thirty fishing sites. We can't just sit here." Even so, in little more than a year after the Spring Bar mining operation had been completed, the huge hole the excavators had dug into the island had already filled in again, with gravel.

III

The Stone People

For Sto:lo people, rocks and stones are involved in events at the beginning of time, and in their own beginnings. There are stones and stone people in accounts of the distant past, the *sxwoxwiyam*, and rocks and stones feature as well in the *sqwelqwel*, the historical narratives of the more recent past. There are stone people very much in the present, too.

You will find one such stone personage seated comfortably in a miniature canoe in the grand foyer of Building 10, a stylish new red Sto:lo Nation building off of Vedder Road, just south of Chilliwack's massive Cottonwood Mall.

His name is T'ixwelátsa. He is the founding ancestor of the Chilliwack Tribe, one of the largest of the ancient political divisions of the Sto:lo people. His story begins in the mountains above Vedder Crossing, at a place on the Chilliwack River near Slesse Creek. Something extraordinary is said to have occurred there, back in the mists of time.

As we know from the ecological sciences and the findings of geology, there was a great deal of upheaval and chaos during the Early Holocene, as the land re-emerged from beneath the great ice sheets that once covered what we now call the Fraser Valley to depths of up to two kilometres. In the Sto:lo telling, these were times when the physical world was unstable and there was disorder and chaos among and between inanimate things and living things. The accounts that derive from the *sxwoxwiyam* assert that even people and animals had not been properly set apart. People, especially, were pretty clueless about how they should behave towards one another.

It was at this time in history that X:als came, from the west. In some stories, X:als is more than one character, and consists of brothers, and perhaps a woman, too. In others, he appears as a single, mythic figure. In any case, X:als moved through the world, changing things, setting things right, and instructing people in proper conduct. It was during his travels in the Sto:lo country that X:als encountered a human being, a man, fishing on the Chilliwack River. The man was T'ixwelátsa, and as he was fishing he was berating his wife for something and otherwise acting badly. So X:als turned him into stone.

T'ixwelátsa in the Sto:lo Nation's Building 10, just south of Chilliwack's Cottonwood Mall. CHARLES CAMPBELL PHOTO

Down through the ages, generations of Chilliwack people cared for the stone T'ixwelátsa, assigning him a place of honour outside the great houses of important Chilliwack families. The statue was venerated as physical evidence of the Chilliwacks' association with their homelands from the beginning of time, and also of that moment in history when X:als moved across the face of the earth and created a world in which people flourished and prospered.

Then came smallpox, wave upon wave of European settlement, and confinement to Indian reserves. The Chilliwacks were reduced to a tiny remnant of their former numbers. There were empty villages everywhere. Then the Potlatch Law of 1884 disrupted the persistence of customary law by prohibiting the ceremonial assignment of duties, entitlements, and property. The Chilliwacks' stone ancestor ended up alone in the ruins of an abandoned village, just east of the Huntington border crossing, south of Abbotsford. It had gotten there as part of a dowry arrangement of sorts, in the marriage of a Chilliwack woman to a Sumas man, during the middle part of the 1800s.

On September 15, 1892, an article appeared in the *Chilliwack Progress* reporting that the Ward brothers, whose farm was situated not far from Vedder Crossing, had come upon "a curiously carved Indian image" on Sumas Prairie. "The image is about four feet high, and weighs about 600 lbs. It is evidently very ancient, and is quite intact, every detail being clearly defined."

A century later, the chief of the Tzeachten band of the Chilliwack Tribe was Herb Joe, who also happened to be the man who had inherited the name T'ixwelátsa. Herb learned that the stone containing the soul of his tribe's ancestor, the statue that had been so long venerated by the Chilliwack people, was to be found in Storage Room 33 of the Burke Museum of Natural History and Culture at the University of Washington, in Seattle. It was known there as a granite statue, Catalogue #152, Accession #190.

All those years, T'ixwelátsa had been gazing out on his confines with the elliptical eyes that are typical of the human face in Northwest Coast art. All alone, but not forgotten. And while it is commonplace to refer to the stone as T'ixwelátsa, the man, the stone is more accurately understood by reference to a thing you can describe in the Halkomelem language as *shxweli*, or *smestíyexw*. Both these terms can be roughly translated into the English as

Hatzic Rock at Xa:ytem, in Mission, believed to represent three Sto:lo leaders turned to stone by the mythic Xa:ls. CHARLES CAMPBELL PHOTO

soul, or "life force." So, more accurately, the Chilliwack elders say, it is the living soul of T'ixwelátsa that resides within that stone statue in the Sto:lo offices on the fringes of Chilliwack's sprawl.

How it came to pass that T'ixwelátsa returned from the Burke Museum to his home place involves a rigmarole that Herb Joe endured through negotiations, research, applications and proceedings under the United States' Native American Graves Protection and Repatriation Act (NAGPRA). Herb set out on his task in 1992, and it took until March 2006 before it was confirmed that T'ixwelátsa was to come home. Weeks after the news of his victory, Herb was still struggling for words. "It's just so overwhelming," said Joe. "I can't say how I feel about it. But it's so exciting. I feel so grateful and fulfilled."

Dave Schaepe, an archeologist who works with the Sto:lo, and who worked on the case with Herb for a decade, said the stone is so unique that it's difficult to draw comparisons from other cultures to describe its significance. Imagine if Moses came down from Mount Sinai, but instead of returning with commandments on a stone tablet, he was turned to stone himself. "It would be a bit like that," Schaepe said.

The place the stone occupies in Chilliwack culture and history is so distinct that the case for repatriation under NAGPRA was made not only under the law's provisions for returning objects of cultural patrimony, but also under NAGPRA regulations governing the disposition of human remains. Further complicating matters, NAGPRA doesn't allow for objects to be directly transferred outside the United States, so the stone had to be handed over to the Chilliwacks' American neighbours in the Nooksack Tribe.

The Nooksacks duly transferred T'ixwelátsa north across the border, and a huge tribal welcoming ceremony was held for the Chilliwacks' ancestor on October 14, 2006 at the Sumas Longhouse, on the Kilgard reserve.

While none is so beautifully and meticulously anthropomorphic as the T'ixwelátsa stone, there are several such "stone people" throughout the Sto:lo territory — another term that that can be used to roughly delineate Sturgeon Reach, the Gravel Reach, or the Heart of the Fraser. The best known of these is Hatzic Rock, at Xa:ytem, within the municipal boundaries of the City of Mission. As the story goes, Xa:ls transformed three Sto:lo leaders here into a single stone. Xa:ytem is an important archeological site, a former village that had been occupied for thousands of years. During the 1990s, a longhouse and interpretive centre were constructed there to promote tourism and education, but they have fallen into disuse.

The point of all this is to make plain that in the conflicts that have emerged over gravel mining in the Fraser River, the Sto:lo point of view is not easily reconciled with environmentalist notions that separate nature from culture, and wilderness from human settlement. Neither does the Sto:lo interest easily conform with the interests of the real estate industry or the gravel-mining industry. Among the Sto:lo, there may well be a widespread inclination to cleave to ancient worldviews that seem anachronistic. But the Sto:lo are also forward-thinking, entrepreneurial, and increasingly possessed of an acute awareness of their rights to exploit any commercial opportunities that may arise within their limited sphere of influence. And those rights include gravel mining.

"Sto:lo" means "river." For the Sto:lo people, the river is not and never was a static thing. It gives life, and it takes life away. It moves through the realms of the animate and the inanimate, the spiritual and the economic, the living and the dead.

Just as the Sto:lo have been marginalized in their own homelands and confined to their reserves along the riverbanks, the Fraser, too, has been steadily confined within the constrictions of dykes, rip-rap barriers, industrial belts, farmland, and subdivisions. There is only so much room within which the Sto:lo and the river can move. Seabird Island was not especially unique. Several Sto:lo reserves lie "outside the dykes" or are otherwise situated in notoriously flood-prone riparian areas. The Seabird Island people, along with the Cheam, Popkum and Skway communites, were all involved in gravel mining. On the Chilliwack River, the Soowhalie people had been mining gravel and building dykes, just to keep from getting washed away during the torrential spring rains, for years.

This didn't mean the Sto:lo were oblivious to the possible implications. Sto:lo Grand Chief Ken Malloway had emerged as an especially cautious voice in the debates, urging Sto:lo communities to be very careful before jumping aboard the gravel-mining bandwagon. "They need to really take a good look if they're doing any damage to habitat and to see if selling our territory by the truckload is the best way to deal with our resources," Malloway said in an newspaper interview. "When the fish are gone, our aboriginal rights are gone. People say they're doing it for flood control, but there's an awful lot of money to be made in gravel. It sure seems to me that that's the main reason to remove gravel — to sell it and make some money at it."

IV

Wandering Channels, Giving and Taking, and A Still-Beating Heart

In black and white aerial photographs, Spring Bar, during low water levels, looks a lot like a marbled pork chop. The vegetated inner island is the dark meat, the outer white shorelines and occasional transects of exposed gravel are the fat. Islands like Spring Bar either build (aggrade) or erode (degrade) depending on dynamic events in the river between Hope in the northeast and Mission in the southwest. While aggrading, the bars are net deposition zones for gravel that's pushed downstream by river currents. Such processes occur gradually over many, many years, before reversing. While degrading, the bars or riverbanks lose gravel as the currents eat away at them. Below Mission, as the river assumes an even gentler descent towards the sea, gravel bars no longer occur.

Not yet, anyway.

To understand how the Gravel Reach of the Fraser became such a valuable ecological niche for salmon, steelhead and sturgeon — and how it became such a flashpoint in the debate over land use in the Fraser Valley — it helps to understand a little about the larger Fraser River system itself.

The Fraser and its many tributaries, including the Quesnel, Chilcotin, Nechako, Bridge, North and South Thompson rivers, constitutes the single largest watershed in British Columbia, draining one quarter of the provincial land base. It's an area roughly the size of California. As the flows of its upper tributaries combine, the main river gathers force, particularly in the canyons that run from the confluence of the Chilcotin south to Yale, 270 kilometres distant.

During the Miocene Period of 23.7 million to 5.3 million years ago, the land on either side of the canyons (now known as the Interior Plateau) uplifted, while the water cut a deeper and deeper channel through them. Today, the river courses through the canyons with brute force before breaking free of its walled confines near Hope to make a dramatic, near-right-angle turn that takes it in a generally southwest direction towards Vancouver, some 150 kilometres away.

Below the bridge at Hope, the Fraser's current can push an astonishing 275,000 cubic feet of water each second. This is enough water to fill three Olympic-size swimming pools to overflowing in a few rapid blinks of the eye. A force that strong is more than enough to roll quite a few stones along with it, which is how the Gravel Reach got all those gravel bars, and its name. Below Mission, as the drop in elevation eases further, the transport and deposition of waterborne silt take on greater importance. On certain days, depending on prevailing winds and tides, the lower Fraser's strong currents are capable of carrying the silt-laden, brown river water over the more dense saltwater of Georgia Strait as far as Mayne and Galiano Islands, twelve nautical miles from where the Fraser meets the sea.

Present-day gravel deposition in the Reach would have been impossible without the dramatic events associated with the great periods of cooling and warming that shaped, and continue to shape, the physical landscape of the region. The last of these great events was at its zenith about 18,000 years ago. At that time the valley lay under a massive sheet of ice, in places nearly two kilometres thick. But only thirty kilometres or so south of Hope, the imposing Cascade mountains blocked the advance of the ice, leaving most of the interior of present-day Washington State an ice-free refuge for animal, fish and plant life.

A turning point came about 14,000 years ago, when average temperatures began to grow markedly warmer. For the next three millennia the bluish white walls of ice retreated north during a prolonged period of melting. A geologically brief, 500-year hiccup known as the Sumas Advance brought a temporary end to the dramatic warming; colder temperatures prevailed, allowing the ice to re-advance to within forty kilometres of present-day Vancouver.

From then on, the region was generally considered to have entered its "postglacial" age. Over the ensuing 10,500 years, that age would come to be divided into three unique phases. The first, generally cool and moist, lasted about 2,000 years, ending around 6500 BCE. The second lasted nearly three times as long and saw average temperatures climb above those of today. Phase three, the "late postglacial," began around 1000 BCE and has prevailed pretty much until present times. The phase has been characterized by generally cooler and moister conditions.

Now, the emerging scientific consensus is that we are on the cusp of a new phase, thanks mainly to rapidly accumulating greenhouse gases in the earth's atmosphere. Partly a consequence of the burning of fossil fuels, the build-up of heat-trapping gases such as carbon dioxide and methane will, it is widely predicted, result in higher average temperatures. It is also expected to lead to more chaotic and unpredictable rainfalls and snowfalls in the Pacific Northwest corner of the continent, which will have consequences for river systems throughout the region.

In the first decade of the new millennium, residents in the Fraser Valley had already got a taste of what that might mean. In 2007, communities along the Gravel Reach braced for what water experts warned was a potentially devastating flood. The proximate cause of the predicted calamity was the swelling of rivers from rapidly melting snow packs in the mountains in the Upper Fraser watershed. If events conspired to produce a prolonged stretch of warm spring weather, all of the melting snow would push water levels higher than the dykes containing the lower river, producing a flood equal to or greater than the historic flood of 1948. But rapidly melting snow packs were only part of the picture.

A more complex, warming-related phenomenon involved the dying of millions of pine trees over an area in the interior of British Columbia roughly the size of England. The lifeless roots of those trees no longer sucked up water and their needleless branches no longer intercepted snow or provided much by way of shade. The pines had been killed in an epic insect attack. A succession of unusually warm winters had allowed tree-killing beetles to mushroom in number. As if the warming-related beetle outbreak was not enough of a threat to downstream environments, a frenzied period of clear-cut logging in response to the beetles complicated matters further still. Hillsides denuded of all their trees allowed snow packs to build to spectacular depths in the winter months and melt all the faster in spring when the sun's rays, unobstructed by tree trunks and branches, beat down upon them.

The floods of 2007 were nowhere near as severe as feared, a matter of luck more than anything. Temperatures simply didn't warm as fast as they could have, nor were the high temperatures sustained. Had they been, the words *global warming* would have featured prominently in news accounts of the calamity.

But back now to earlier times.

What were the consequences of a rapidly warming world on the Lower Fraser and its environs as the last Ice Age came to an end? The most obvious thing was that all of the standing ice that had exerted such tremendous pressure on the earth below started disappearing As massive ice walls melted and the glaciers receded, the ground underneath, which had been lowered 300 metres and more, rebounded. By 9500 BC the land surface had come to

Looking upriver from Chilliwack Mountain. BARRY STEWART PHOTO

within about thirty metres of the elevation it's found today.[9] The rising earth, which reappeared from under ice and water, allowed the river channel to cut deeper and deeper into it. The other thing that the glaciers did as they advanced and retreated was to act as super-powerful scours.

"Vast amounts of boulders, gravel and dust were created and distributed by the immense moving ice sheets and were then left in the wake of the glaciers," scientists Marvin Rosenau and Mark Angelo wrote in a 2000 report detailing the importance of sand and gravel in salmon, steelhead and sturgeon streams in British Columbia.[10] Into this environment, fish populations that had survived the prolonged period of colder weather in ice-free refuges — some in present-day Alaska, some as far away as Mexico — slowly began to colonize new habitat in the Fraser system. "Many of the stream bottoms and valleys constructed by the retreating ice became storehouses of clay, silt, sand, gravel cobbles and boulders, and these are the parent components of modern-day salmon and steelhead watersheds."

The operative word here is *slowly*. For one thing, much of the present-day lower river and its tributaries didn't exist 11,000 years ago. Still, the earliest evidence of human habitation in the region was uncovered in the southern reaches of the Fraser Canyon — the Milliken site — and dates back more than 9,000 years. Back then, the Fraser Valley was essentially a long fjord that tapered to a head near present-day Yale.[11] Salmon populations first re-established in and around the Yale area, and with them, human settlement. Significantly, the earliest human inhabitants in this new, ice-free landscape used the cobbles and rock left behind by the glaciers to chip away at and form rock tools.

As the upwelling of lands freed from ice continued, the rough outlines of the present-day Fraser Valley emerged. Powerful river and stream currents fed by the retreating ice carved new and continually altering courses through the rock and sediment left behind by the glaciers. The prevailing environment remained harsh. While no longer buried beneath ice or water, the landscape was largely bereft of vegetation. It would take thousands more years for soils to sufficiently build on the riverbanks and hillsides to support a rich tapestry of plant life, including the tall cottonwoods of the floodplain and the dense evergreen cedars and other

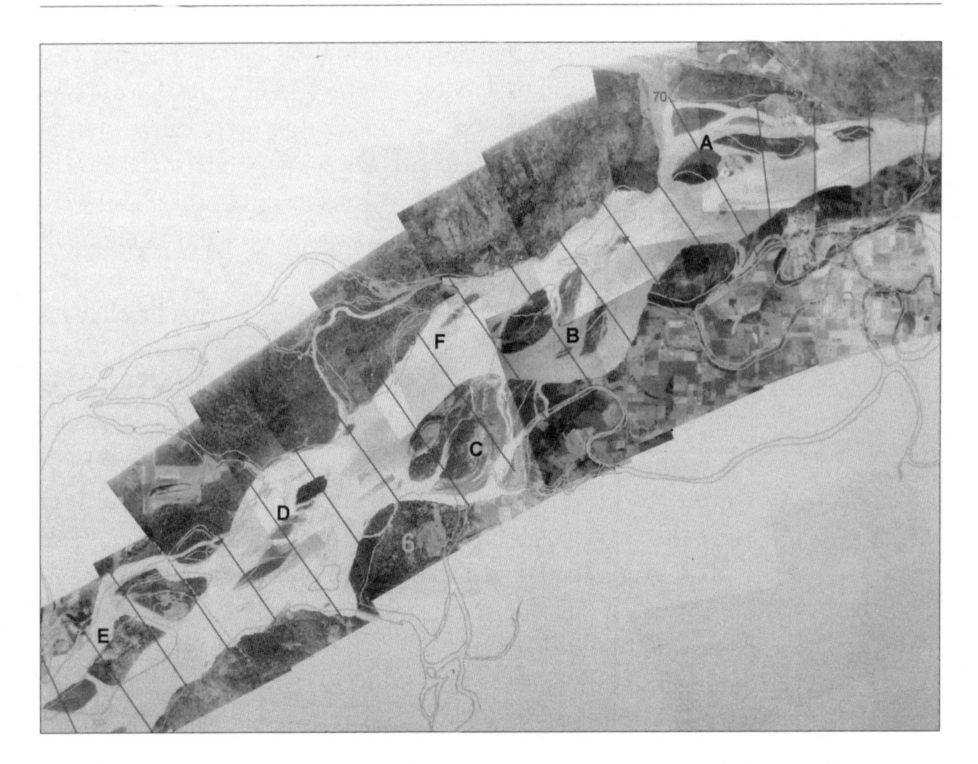

By comparing composite aerial photographs like this one, compiled from photos taken of the Fraser River on July 15, 1928, with photos of the same stretch of river taken in later years, Michael Church and his colleagues are able to track the build-up of sediment and the formation of bars and islands. The lines in the above compare 1928 and 1999 morphological features of the rivercourse. Harrison Knob is near the top right of the photo, near the numeral "70". CHURCH & HAM (2004)..

conifers that marched up the surrounding mountains. Only with vegetation firmly established some 4,000 years or so ago, did the low-lying valley lands and waterways cutting through them stabilize and assume their central importance as one of the richest repositories of fish life in North America, and a food source critical to the First Nations people who harvested them sustainably for millennia prior to the arrival of European settlers in the 1800s.

Stability, however, was always a fluid thing in the Gravel Reach. In the intervening 4,000 years, the landscape of this stretch of the river constantly shifted. The shifts explain both how places like Spring Bar arise and how they subsequently disappear to be replaced somewhere else by new bars rising from the river's depths.

The dynamic events that give birth to these bars and islands also raise serious questions about government policy that encourages gravel extraction operations of the sort that were carried out at Spring Bar. Over the space of several weeks in winter and spring of 2007 and 2008, more than 30,000 truckloads of river gravel were removed from the bar — ostensibly to reduce flood risks but in all likelihood doing nothing of the kind.

Gravel bars emerge and disappear because of dynamic changes in the river's behaviour as it leaves one landscape behind to enter a dramatically different one. Michael Church, a University of British Columbia geographer and expert on what's known as "the sediment budget" in the Gravel Reach, says it's the Fraser River's sharply reduced gradient as it exits the narrow canyons north of Hope that is the major cause of gravel deposition downriver.

"The river follows a steep, confined course through the mountains where it picks up rock, gravel, sand, silt and clay from the banks and from tributaries. Within the Lower Mainland the gradient of the river declines quickly as it approaches the sea. It cannot continue to move the larger material on the reduced gradient. The largest material is abandoned first, so that the river between Hope and Sumas Mountain flows over its own gravel deposits."[12]

This is typical of rivers that exit mountainous regions and encounter lowlands. Upon doing so they commonly form alluvial fans or cone-shaped deposits of river-borne gravel and sediment. These fans continually accumulate sediment because the rivers cannot carry gravel across them and beyond. In the Fraser River's case, this means that in the river to the east of Mission, the Gravel Reach is generally aggrading, or raising its bed each year, although, very importantly, it remains far from a neat, steady accumulation.[13]

Rivers moving through such environments tend to meander, particularly when they are unconstrained by natural or man-made barriers such as dykes. Currents erode the riverbed and move laterally until they encounter a hard valley wall or some other obstruction. This triggers a reversal, and erosion then moves in the opposite direction to another unmovable obstruction, and things reverse again. Over tens, hundreds or even thousands of years the pattern repeats. The end result is a continuously sinuous river channel.[14]

In the Gravel Reach, this lateral movement was always somewhat more constrained because of the nearby mountains. Nevertheless, it remained in Church's words a "wandering channel". This wandering gave rise at places to the braiding of the river into parallel channels, some of which were year-round secondary channels to the main river, and others were more minor and, in low-water months, completely dry. This wandering action is also what gives rise to gravel bars and, eventually in some cases, islands.

Gravel depositions vary tremendously in volume and are usually offset to a significant degree by erosion somewhere further downstream. It is this give and take that shifts the river's bars and islands. Typically, as gravel is carried, pushed or rolled downstream by the river current it is eventually deposited at some point where the current loses its velocity or where the water flow is forced to diverge around some obstruction. "In such places," Church writes, "material accumulates to form the major bars that redirect the flow."[15]

If the buildup occurs in the middle of the river, it may split the flow in two. This may then result in erosion on both riverbanks opposite the bar. Where a bar forms predominantly on one side of the river, the current may push off of it to "initiate attack" on the opposite bank. Such attacks loosen gravel, which is then transported downstream and dropped at another point where deposits build again. Predictably, as such deposits build, new attacks are initiated on other bank sections or bars further downstream.

For many years, areas of the Gravel Reach may be either persistent points of deposition or erosion. The more persistent gravel bars may ultimately form layers of waterborne silt on top of them and later thick mats of vegetation including tall cottonwood trees. As bars slowly build or erode, dramatic realignments eventually occur. These include shifts in the main river channel itself and the disappearance of certain bars and islands. But the operative word here is eventually. "The reason why it takes years to 'reorganize' the channel and change the local channel activity is that the river channel is large, whilst the total load of bed material transported down the river remains modest," Church notes.[16]

"Give and take" events in the Gravel Reach are central to understanding why it provides such exceptional habitat for fish. Because gravel is routinely moved along the riverbed, typically during the May to August freshet period, it does not become compacted and maintains a relatively loose character. This allows the salmon that spawn in the Gravel Reach to more easily use their tails to dig holes in the gravel. The holes, known as redds, are where eggs are laid and then fertilized. Once this is done, the salmon set about backfilling the holes with gravel again. During the incubation period, the water flow constantly cleanses the porous gravel, keeping the redds free of fine waterborne sediments that would smother the eggs. When a new generation of salmon is born, the gravel provides valuable protective cover and later rearing habitat and migration corridors for the vulnerable young fish.

Secondary channels and slack water areas in the Gravel Reach also provide lengthy shore banks along which the young fish can hide and feed. Varying depths of water off the gravel banks also provide an optimum range of habitats for the young salmon as well as for the numerous invertebrates they eat. Elsewhere, "scour holes" created where the strong river current directly hits gravel bars or islands provide perfect deeper water habitat for larger adult fish, including sturgeon. Finally, the varying size of gravels and sediments encourage the widest possible array of habitats for young fish as they grow in size.

Put together, all of these things help to explain why more than thirty different species of fish are found in the Gravel Reach. Such diversity is made all the more remarkable when one considers that by the dawn of the new millennium, more than two million people were calling the Lower Fraser and environs their home, and for more than 150 years the river valley west of Hope had been the site of an explosive growth in human settlement, a proliferation of farms, numerous mining and logging operations, and all manner of industrial developments. Each of these things has had repercussions for the lower river. Yet through them all, the Fraser's heart has kept beating.

V

Memories of the Great Flood

The Gravel Reach is inextricably linked to human habitation in the region dating back 10,000 years, and by the late twentieth and early twenty-first centuries, archaeological and anthropological studies had confirmed much of what aboriginal oral histories had long maintained. Meanwhile, ongoing studies by biologists confirmed and continued to refine human understanding of both the diversity of fish species in the Gravel Reach and their unique and varied habitat needs.

All of this amassed knowledge made it difficult to approve industrial-scale mining activities. The gravel was undoubtedly there in abundance, but the ecological consequences of removing it in any quantity, without robust planning and risk assessment, were dire. So, if commercial exploitation of the lower river's rich gravel bars and islands were to take place, it had to be justified in the name of some higher purpose.

That higher purpose became the emotionally charged issue of flood control.

While the over-hyped catastrophic "flood" of 2007 failed to materialize, the Fraser did overrun its banks to devastating effect in decades past, most notably in 1894 and 1948. The realities of those events are not lost to the Fraser Valley's farmers.

High snow packs in the upper tributaries of the greater watershed, followed by an intense and sustained period of springtime heat, were the proximate causes of the 1894 flood. Because the flood occurred at a time when human settlement was far more sparse than in subsequent decades, it had a minimal impact on homes and businesses. But it was a spectacular and terrifying event nonetheless. Floodwaters inundated extensive areas of land from the confluence of the Harrison with the Fraser near Chilliwack all the way to the river's delta lands in present-day Richmond. A subsequent series of hydraulic investigations more than a century later would conclude through modelling work that the peak flood discharge rate at the height of the flood sent an astonishing 17,000 cubic metres of water per second past Hope as the river exited the upstream canyon.[17] By comparison, more contemporary analysis of water flows at Hope place low-water flows averaged over a ten-year period at 450 cubic metres per second, and mean annual flows at 2,700 cubic metres per second.

The 1948 flood was not quite as severe, although nearly so. However, in the intervening 54 years, the Fraser Valley's population had grown considerably. When the river overflowed its banks, the ensuing damage to farm buildings, crops and forage lands, homes and businesses was massive. The rising waters forced the evacuation of 16,000 people, damaged or completely destroyed 2,300 homes and left 1,500 people homeless.[18]

Because of the damage, the 1948 flood was the more important of the two historic events. The Great Flood of '48 unleashed a torrent of flood-control infrastructure investments. By the end of the twentieth century, an extensive network of newly constructed or fortified dykes stretched some 600 kilometres had been constructed, along with 100 pump stations and 400 flood-control boxes.[19] All of this was designed to try and prevent the river from overrunning its banks and damaging an expanding number of homes, businesses and infrastructure, including water and sewer mains, roads, bridges and rail lines. And all of this new infrastructure required gravel.

From 1970 onward, the provincial and federal governments combined to spend approximately $300 million on upgrades to the flood control system through the Fraser River Flood Control Program. Further efforts were made to "armour" riverbanks by lining them with riprap. Fortifying the banks with rock was intended to prevent the river from eating away at one bank while simultaneously building a bank, bar or island somewhere else through waterborne gravel deposition.

The 1948 flood, near Chilliwack. CHILLIWACK MUSEUM AND ARCHIVES

Provincial and federal regulators continued to maintain that the highly contentious practice of removing gravel from the river's bars was necessary to reduce flood risks. The known threats such actions posed to the river's delicately balanced ecology were put aside on assumptions about the Gravel Reach as a net deposition zone for gravel. The idea was that removing gravel commensurate with what the river annually deposited would keep water levels in some state of balance.

This sounds like a good idea, but making it work requires a fairly accurate reading of how much upriver gravel is naturally deposited in the Gravel Reach in the first place. Michael Church, the UBC sediment-budget expert, reckons that the federal and provincial governments have overestimated natural gravel-deposition rates in a way that favours arguments for inordinately high rates of gravel removal.

Church says it's also likely that the river's vast gravel deposits are not in fact a wholly "natural" phenomenon at all. He points to the rarely considered impact of late nineteenth century hydraulic placer-mining operations in the Fraser Canyon, which dredged and blasted huge volumes of gravel in a low-technology gold rush.

In any case, by the first years of the twenty-first century, a new gold rush was in the works. Following a five-year moratorium that was put in place to protect fish habitat, the federal Department of Fisheries and Oceans and the BC government signed a five-year Letter of Understanding, in 2004, to proceed with gravel mining in the Heart of the Fraser. While earlier efforts had mainly aimed at the river's main channel to ease passage for boats, tugs and barges, the new deal was of a different scale entirely. It allowed companies to excavate a million cubic metres of gravel from the Reach between 2005 and 2006, and a further 1.26 million cubic metres between 2007 and 2009.

The Spring Bar operation was a big part of that effort. The public was starting to notice, and conservationist concerns were met with hyperbolic justifications. John Les, then British Columbia's Minister of Public Safety and the MLA for Chilliwack, said: "Our sole motivation for removing gravel is flood protection. At the end of the day, we don't want to be playing Russian roulette with the safety of citizens in the Fraser Valley and Lower Mainland."[20] The comment was music to the gravel industry's ears, and was promptly reprinted in the industry publication, *Dredging News*.

Implicit in the new federal-provincial pact was that it was perhaps okay to play Russian roulette with fish, which is precisely what had happened in 2006 when roughly 54,000 cubic metres of gravel was removed from the a massive gravel deposit called Big Bar, near Rosedale, just downstream from Seabird Island. The Big Bar operation presaged the Spring Bar controversy. In the Big Bar case, the Cheam First Nation partnered with a local gravel company, and things didn't exactly work out as planned. Somewhere between 1.5 and 2.25 million pink salmon hatchlings died at Big Bar.

To access the bar in the middle of the river, the gravel company built a temporary causeway — essentially a gravel road over a section of the riverbed. Constructed from the Fraser's south bank just downstream of the Agassiz-Rosedale bridge, the temporary road proved fine for heavy equipment, but disastrous for fish. The channel downstream of the bridge unexpectedly dried up, stranding salmon fry as their water-bound gravel homes disappeared.[21]

The events at Big Bar forced federal fisheries officials to make the understated admission that the mining operation had not gone "as smoothly as anticipated".[22] The Big Bar debacle was also the trigger that heightened anxieties when the Spring Bar operation was about to begin. In a complaint filed with Canada's Auditor General, critics of the Spring Bar removal, including the outspoken former provincial biologist Marvin Rosenau, noted that the proposed extraction of 400,000 cubic metres of gravel would constitute the largest removal of its kind from a single bar or island in the lower Fraser River since statistics on gravel mining had first been compiled.

The unprecedented operation resulted in an excavation over a combined area roughly 150 metres wide, up to five metres deep and nearly a kilometre long. The size of the mining operation meant that at least ten hectares of prime habitat for rearing Chinook salmon was destroyed. But the mining of Spring Bar presented other threats as well. In the right channel opposite the bar, pink salmon were known to spawn in abundance. In the event that mining operations in the left channel caused significant changes in water flow volumes, it was possible in subsequent years that the right channel at Spring Bar could "de-water," causing significant collateral damage.

The justification for the Spring Bar operation was that the area was generally filling with so much water-transported gravel that the river was at serious risk of flooding. Yet, an exhaustive report on site-specific build-ups and erosion of gravel throughout the Gravel Reach between 1952 and 1999 showed that nearly 4.1 million cubic metres of gravel had eroded from the Spring Bar stretch of the river. And as for Spring Bar itself, at least half a million cubic metres of gravel had eroded from it during that timeframe.[23] Far from building with gravel and raising water levels, the Fraser, over the course of nearly half a century, had significantly eroded riverbanks and bars in that stretch of the river. The area was not imminently threatened by rising water levels—quite the contrary. If a risk existed, and it was linked to gravel deposition, it lay somewhere else.

Through decades of the study of gravel transport in the Gravel Reach, Michael Church had come to believe that some amount of gravel extraction should occur. But where, when and how were critical to ensuring the least possible ecological harm. His work suggested that on average the net deposition of gravel in the Reach

Michael Church in his UBC office. CHARLES CAMPBELL PHOTO

was somewhere between 200,000 and 300,000 cubic metres per year; a seemingly large volume of rock, but infinitesimally small when viewed against a hundred-kilometre stretch of river and a main river channel spanning one half kilometre.

As controversy around the Spring Bar removal swirled in the winter of 2008, Church argued no more gravel than was annually deposited in the Gravel Reach should be removed in any given year. The Spring Bar gravel removal alone was potentially double what the river would deposit along a hundred-kilometre stretch in a full year. The Spring Bar mining operation hit one site and one site only— a site that was naturally eroding, not building with gravel.

"In most rivers of the world where gravel has been removed it's been removed for commercial purposes . . . at rates greater than the annual input rate, and what we universally see is the collapse of the river into a single thread, [a] deep, barren channel," Church said.[24] Furthermore, targeting one bar for such an extensive mining operation would not necessarily improve flood safety, because doing so encouraged the river to flow in different ways leading to more erosion in some places and greater gravel deposition in others downstream of the mining operation. In other words, it had the potential to increase site-specific flood threats downstream.[25]

What then of the entire five-year program of gravel removal? Would pulling 2.26 million cubic metres of gravel from the Reach's bars and islands really reduce flood risks? A study commissioned by federal fisheries officials and published at the mid-point of the controversial mining program said no. The research, by professional engineers concluded that even if that massive amount of rock was pulled from the Reach in one fell swoop, it would have a negligible effect on reducing flooding. Indeed, water flow models developed by the engineers suggested that in the Gravel Reach immediately downstream of where the Harrison River met the Fraser River, the drop in water levels achieved by removing all that gravel would be in the order of one inch. Upstream of the Harrison confluence, the drop would be a somewhat larger two to four inches, for reasons having to do with the unique topography at the confluence of the two rivers. Viewed against the Fraser's potential to push huge volumes of water through at peak flows, however, such drops were inconsequential.

The engineer's blunt assessment: "It does not appear that large-scale gravel removals from the gravel reach of the Fraser River are effective in lowering flood profile."

It is on this point that Seabird Island's Clem Seymour, in his cramped and cluttered chief's office at the band council complex on the reserve, throws up his hands in exasperation. It is all very well and good, these interesting calculations, he said. But in the meantime, in the engineer's report's exhaustively documented case against the notion of a great flood threat, the erosion so assiduously estimated in the Spring Bar section of the river was occurring at the expense of the Seabird Island Indian reserve.

While Seymour's claims are hotly contested, it is in these ways that non-aboriginal conservationists and local aboriginal people have sometimes found themselves talking over one another's heads, confusing one another's arguments and mistaking one another's entreaties, from one end of the Gravel Reach to the other. You could say it's a sort of tradition. It goes back to the very first conversation the Sto:lo ever had with the newcomers, which also involved rocks and stones.

VI

'Gritting Its Teeth' Rock

At one of his last encounters in the realm of the river people, before he went up into the sky to become a star, X:als is said to have engaged in a great battle with some sort of shaman, about 120 kilometers upriver from the sea, at the upper extent of Sturgeon Reach. It happened at a place within the steep walls of the Fraser Canyon. The place is called Tatxlis, which means Gritting His Teeth. The shaman, whose name was Qewxtelemos, stood on the far bank of the river, and he and X:als battled with strange medicine powers and thunderbolts. There is a large stone at Tatxlis, and it is covered in deeply indented striations, like scratch marks, which were made, they say, when X:als worried his thumbs into the rock.

In the summer of 1808, as the great explorer Simon Fraser was nearing the close of his five-year journey from Montreal, he came upon many people and many stout villages. He was brought to the Tatxlis stone, and its adjacent village, on the afternoon of June 28. That evening, he recorded this in his diary: "At the bad rock [Lady Franklin Rock], a little distance above the village, where the rapids terminate, the natives informed us that white people like us came there from below; and they shewed us indented marks which the white people made upon the rocks, but which, by the bye, seemed to us to be natural marks."

What appears to have occurred here is that the Sto:lo people of the canyon mistook Fraser for a relation of X:als, the Transformer, and Fraser mistook his Sto:lo hosts to mean that white people made the marks in the stone.

It was also about the time that Fraser was coming out of the canyon that he came to realize that for five long years, he had been wrong. The entire purpose of his journey was to chart the great River of the West from the headwaters to the sea. It was only in the canyon, in the days before his final descent to the sea, that he realized he'd been mistaken from the start. This was not the Columbia River. This was another river, and it was now turning west, towards the seacoast, several degrees of latitude north of where the Columbia mouth was known to be.

At Gritting Its Teeth, the Sto:lo tribal historian Sonny McHalsie can show you exactly the place that Simon Fraser was brought in 1808, at the very stone where X:als is said to have worried his

A crop of river rocks — one of the products of the gravel trade. CHARLES CAMPBELL PHOTO

thumbprints into the very rock. Generations of archeologists have come to this place. They have scoured and mapped and carefully examined the terrain and they have noted the precise location of almost every fire-cracked rock, every hearth, and every ancient post hole, and they have assigned protected-site archeological registration numbers to each of them.

But there was something all those archeologists never saw, even though it was right there, in plain view, a mere 100 metres or so from the Gritting Its Teeth Rock, on the far side of the river, at a place called Xelhalh. It is the remnant of a massive stone wall. What's still left standing is a section about 20 metres in length and more than six metres in height. There are other walls just upriver of Xelhalh, only much bigger.

The only reason that government archeologists never recorded these strange structures — not until Sto:lo elders and archeologists pointed them out and did the spadework, anyway — was that it was simply presumed that "Indians" in this part of the world did not build things from rocks.

The walls were apparently built as defensive fortifications against the depredations of the enemies of the Sto:lo, warriors from the coast. Sto:lo researchers reckon that the network of fortifications were situated in such a way as to allow coordinated defensive operations. The Sto:lo used a kind of slingshot, too, to hurl rocks at their tormentors.

VII

What Happens at Harrison Knob

The Harrison River is the most important tributary to the Fraser downstream of Hope. Its waters enter the Fraser on the Fraser's north bank and are capable of increasing the main river's flow by a tenth. Even more important, at this confluence, downstream of Seabird, there is a wall of rock and steeply sloped cone of land known as Harrison Knob. Situated at the mouth of the Harrison on its west side, Harrison Knob forces the Fraser to make an abrupt 90-degree turn to the south before the river resumes its generally southwestward direction. It helps to understand how that big, intimidating chunk of land influences events in the Gravel Reach by thinking about what happens when you are barreling down the highway in your car and suddenly realize that the cars and trucks ahead of you are braking because there's a sharp turn ahead. You apply your brakes and the reaction continues through the traffic behind you. Within seconds, you are part of a long and rapidly lengthening line of cars, trucks, buses, motorcycles and the like. You are in a traffic jam.

Something very much like that occurs as the Fraser's flow, increased by the Harrison's, encounters the Knob. Suddenly, the river's forward momentum is blunted and a water jam of a sort is the result.

Robert Millar, a professor of engineering at the University of British Columbia and an expert on the river's flood profile, explains: "There's an energy loss. From a lay perspective it causes the river water to back up. Imagine a four-lane highway essentially having to make a 90-degree turn. Just like the traffic backs up, the water backs up." And, as the water backs up, upstream of the Knob, "it gets higher."[26] How high depends on the flow. High flows mean higher backups. But whether low or high, the result is a backup, triggered by the Knob. It is most pronounced just upstream of the partial obstruction and dissipates further upriver, finally disappearing about seven kilometres from where the Harrison joins with the Fraser.

Significantly, things were not always this way. In the recent past the Fraser River did not hit with nearly such force on Harrison Knob, and there was nowhere near the backup. Indeed, aerial photographs of the river from March 1949 show that the main river channel flowed considerably south of the Knob and below a collection of gravel islands around the Harrison mouth and downstream. By March of 1971, the view from above revealed a very different picture. The islands at and immediately downstream of the confluence of the two rivers had amassed more gravel due to the erosion of islands and gravel bars further upstream. The result was that the southern channel had contorted and narrowed, while the northern channel gained more flow and came into contact with Harrison Knob. By 1983, it was clear that the main river channel had shifted to the north, with more of the Fraser's flow hitting the Knob. The next two sets of aerial photographs taken at the beginning and the end of the 1990s, reveal that the gravel islands had grown larger and were covered in more vegetation to the point where the old southern channel, known as Minto Channel, was largely cut off.[27]

"It is conceivable that Minto Channel could again become the main channel in this region of the river," Church concluded in a report to the District of Chilliwack in 1998. Such an event, he warned, could cause the river to erode the left bank and potentially threaten the dykes that protect valuable property in and around Chilliwack."[28]

But there was a more pressing question for others who had studied the region. What are the implications of preventing the river from doing what it would otherwise do, of its own accord?

"If the Fraser were to reoccupy Minto Channel, or cut a new, more efficient channel across Minto Island, flood levels would almost certainly decrease significantly," Millar said. He added that such an outcome had been "very poorly presented" to residents along that stretch of the Gravel Reach. Instead, Russian roulette-like pronouncements that so pleased the gravel mining had created a heightened sense of crisis and a public desire that something should be done before the next freshet. Always, it seemed, that "something" meant gravel mining.

"Gravel removal may play a role in the long-term management of the Gravel Reach," Millar said, "but it is of little consequence in addressing current deficiencies in the dykes. If we are to provide a higher level of flood protection with the current Fraser River alignment, then, in my view, the only real option is to raise the dykes."

VIII

Arguments That Don't Hold Water, Arguments That Do

The 1948 flood forever cemented the idea that engineering works in the Fraser Valley were required to deal with the river's propensity to occasionally swell with dangerously high levels of rainwater and snowmelt. But by the latter part of the twentieth century, it had become apparent that engineering solutions also created as many problems as they solved. An extensive network of dykes had made it possible for relatively safe and prosperous farming to flourish on the rich, silt-laden lowlands of the valley — the most productive farmlands in all of British Columbia, on the doorstep of the largest population concentration in the province. But the dykes also tightened a cordon on the river and gave it a narrower band of land in which to wander. Similarly, lining lengthy areas of shoreline with riprap prevented the river from eroding lands that were subsequently cultivated, foraged or settled. But, paradoxically, the engineering works made the river hit with greater force on other areas downstream, causing more severe erosion in some places and more substantial build-up of gravel in other places. In turn, these events led to calls for more gravel removal, even though the removals had a negligible effect on reducing flood risks.

As "sustainable development" gained currency as an idea in the 1980s, however, it came to be seen that human intervention in natural processes had to be orchestrated differently in order to ensure the persistence of reasonably healthy and diverse ecosystems. Problematically, in British Columbia the debate over what sustainable development meant and how and where it should apply had come to focus on "wild" or "pristine" landscapes and seascapes — areas seemingly untouched by man, even though First Nations people had in many cases lived in and altered natural processes in such places for thousands of years.

In the process, environmental organizations attracted untold millions of dollars, from well-endowed US foundations in particular, to assist them in protecting such areas. The success of such campaigns — seen in the myriad parks created through the 1980s, 1990s and first decade of the new millennium — made it more and more difficult for people to see that some of the richest, most ecologically diverse landscapes were literally in their backyards, in the midst or on the periphery of areas of the longest-standing human occupation.

The challenge in such areas was not to try to turn the clock back to some arbitrarily chosen idyllic time in the distant or not-so-distant past, but to shift human uses of natural resources, in some cases slightly, in others, substantially, to achieve the healthy balance that sustainable development demanded. In the case of the Gravel Reach, in the early years of the twenty-first century, it remained to be seen what that meant. But if some semblance of a healthy river system was to be maintained on the doorstep of the largest human population in the province, it appeared that certain concepts needed to be embraced, and quickly.

It seems sensible to raise and fortify dykes to ensure public health and safety. But it makes less sense to build new dykes and place more riprap alongside a river that is already mostly confined within a man-made straitjacket. First, because of the dire consequences such actions posed to fisheries. Second, because further constraining the river actually increases the intensity of erosion and gravel deposition further downstream.

It is also without question that some engineering in the river itself is a reasonable proposition. If certain stretches of the Gravel Reach become so filled with gravel or sand that they impede boat traffic, then occasionally they might need to be dredged. Some amount of gravel removal might also be justified on grounds that it would demonstrably lower water levels. For example, cutting a trench through the gravel bars south of Harrison Knob, while clearly posing risks to fisheries habitat, might allow the river to bypass a large, immovable object that caused upstream water lev-

Downriver projects, like the Golden Ears Bridge, create massive upriver demand for the Fraser's gravel. ANTHONY DYKE PHOTO

els to rise. But continuing to target individual bars and islands for large-scale gravel excavation is increasingly foolhardy. It has killed fish, and threatens to render extensive stretches of the river unusable as spawning habitat for years if not decades to come, and it has no appreciable impact on reducing flood risk.

Communities along the Gravel Reach will continue to grow and continue to need gravel. But if it came to be seen that extracting gravel from the river to reduce flood control was a false promise, then the deck was cleared for a much-needed discussion about where to find the gravel to build the region's homes, businesses and infrastructure. If not from the Reach itself, where? This led to other questions. How should the gravel be moved? How should it be used?

The sharp rise in oil and gas prices in the early months of 2008, just as the excavators were hauling out gravel at Spring Bar, helped to highlight what everyone already knows. The more non-renewable resources are used up, the more prices climb. Gravel is clearly one of those non-renewable resources. Like other natural resources, prices climb the further from market one has to go to get it. In the case of gravel, however, in the absence of exhorbitant subsidies, just the immense weight of the commodity places limits on how far it can be moved economically.

The Fraser itself can make it pretty cheap to transport gravel. KEN HALL PHOTO

Much of the push to utilize the Lower Fraser's gravel arises from its mere proximity, which makes it relatively inexpensive to move around. As loose gravel supplies on valley lands become less available, the temptation to utilize more river gravel grows. As the twenty-first century began, controversy over the environmental consequences of in-river removals was, for a time, blunted by the assertions of developers and local politicians who insisted that the removals served the higher purpose of protecting private property from floods. But the more scientific studies showed such arguments didn't hold water, the more attention came to focus on where future, less environmentally contentious supplies of gravel or crushed rock might come from.

A large quarry operation above the Gravel Reach itself seemed to offer part of the answer. In the early years of the new millennium, the Mainland Sand and Gravel operation at Sumas Mountain was the region's top rock producer. Each day, trucks laden with masses of material moved from the quarry to barges moored in the river. Once their holds were filled, the barges were towed out into the main channel of the Gravel Reach for the long haul downriver to cement plants near the mouth of the river or for use in a wide array of construction projects. Each year, Mainland blasted another 1.5

million tonnes or so of rock out of the north slopes of the mountain, requiring 600 barge trips or more for downriver transport. Water transport was a key ingredient in the quarry's ongoing success because it allowed large volumes of rock to be moved long distances at relatively low cost. The same volume of rock moved by road to Vancouver would have required 75,000 truckloads.

The corollary of the Mainland story was that large volumes of rock or rock byproducts could also be moved in the opposite direction. If gravel from the Sunshine Coast and Vancouver Island can be shipped to Hawaii or San Francisco, why not move gravel upriver from the mouth of the Fraser? Make use of existing or new offloading sites at key points along the riverbank in the Fraser Valley. Truck the material in short hauls from the offload points to new housing developments or road repair sites. As the Fraser Valley wrestled with the implications of a steady rise in population, such adjustments were clearly needed. From 1960 until roughly 2000, the valley's population had essentially doubled every twenty years.[29] In 2008, the chief statistical agency for British Columbia reported that while that rate of growth might not be sustained in future decades it would, nonetheless, continue to climb significantly. It projected that from a base of approximately 270,000 people in 2008, a valley population of somewhere around 400,000 would be reached by 2036 — a 49 per cent increase in less than 30 years.[30]

BC Stats also reported that over the same timeframe, the number of households in the valley was expected to rise to more than 161,000 from just over 99,000 — a 63 per cent increase.[31] The fact that the projected growth rate for the number of households exceeded the population rate was due largely to a steadily aging population, with older couples and singles expected to live apart from their children in apartments, condominiums or retirement homes.[32] Of course, shifts in government policy could influence how many buildings are required to house people. Financial inducements that encourage the building of suites in existing houses so that older people could live at home with their adult children might, for example, lessen new housing demand.[33] But without such policy interventions or dramatic changes in demographic and migration trends, the agency's projections strongly suggest continued pressure on land and resources in the Fraser Valley.

As BC Stats noted, the number of residential building permits issued between 2000 and 2007 rose from 696 in 2000 to a high of 2,639 in 2006, before tapering off ever so slightly the following year.[34] And at least some of the gravel used to build more than 14,500 residential units during that eight-year timeframe had been conveniently scooped from the nearby river.

Aware of the challenges such developments posed to the maintenance of a "high quality of life," the regional governing body in the Fraser Valley released a growth strategy in 2004 that emphasized sustainable development. Among the strategy's priorities were "better recognition and protection" of the region's natural environment, including "air quality, water resources, fish, wildlife and natural habitat."[35] The strategy also included specific recommendations to protect farmland, recognizing the important economic contribution farming made to the local economy and to provincial food production. Protecting the environment and farmland implied not only that housing and business developments be contained as much as possible within the boundaries of the six major municipalities within the Fraser Valley, but also that there be a commitment to increased housing densities and more compact housing. The strategy also specifically recommended that new settlements be built in ways that minimized the "risks associated with flooding."[36]

The latter, in particular, was important insofar as the Gravel Reach was concerned. Because if new housing and business developments were ruled off-limits on the floodplain, and in particular on the river side of any dykes, then there would be less call to remove gravel from the river and less pressure to build on valley-bottom farmlands.

Such strategies have almost always been long on principles and short on details. But details obviously loomed large in any regional growth plan, because regardless of where buildings and related infrastructure were located, they would still require the extensive use of materials. Some of the stuff would come from renewable resources. Others would not.

IX

The Concrete Facts

Greg Johnson, a professor of architecture at the University of British Columbia, began to focus on construction methods and materials with an emphasis on sustainable energy use in the early 1980s. By the time the controversy at Spring Bar erupted, he had devoted most of his working life as an architect and teacher to the subject. He had come to see that while building materials such as concrete were highly desirable from the perspective of longevity, they had their problems.

Concrete's big drawback has always been the energy required to make it and to transport it. For example, it requires nearly three times the energy and therefore three times the greenhouse gas emissions to produce the equivalent amount of concrete slabs as it does softwood lumber.[37] Moreover, the raw material to make lumber — trees — can be regrown, and fairly quickly. As they grow, trees pull heat-trapping carbon dioxide out of the atmosphere. When logged, some of that CO^2 will be quickly released. But the locked-up carbon stored in the lumber made from those trees can, with proper attention to architectural design and construction, remain locked up for decades, if not a century or more. Gravel or crushed rock, the critical component in concrete, has no similar benefit.

If anything, however, concrete had been ascendant as a building product, at least in the Greater Vancouver Regional District to the west of the Fraser Valley. There, the fallout from years of bad publicity due to rainwater damage to hundreds of wood-frame

The gravel yard on Seabird Island. CHARLES CAMPBELL PHOTO

condominium and townhouse structures had turned many people against wood and in favour of concrete, glass and steel, Johnson said.[38]

What concrete continued to suffer from, however, were the high costs associated with making it, transporting it, and disposing of the stuff when old buildings are knocked down.

High energy costs were, however, leading to innovations. As one of the architects involved in building a new French-language school in Esquimalt near Victoria, Johnson had overseen a project that included re-using all of the broken-up concrete from the old building that previously stood on the site. A crusher was used to break up large chunks of concrete into much smaller pieces. Such "rock" could not be reused to make concrete because of its chemical properties. But it was perfectly usable for fill or to line perimeter drain lines, or as an underlying layer for things like parking lots. By doing these things, Johnson said, the builders at L'École Victor Brodeur School saved having to truck tonnes of new gravel onto site as well as having to truck tonnes of broken-up concrete off-site. There was no reason, in Johnson's estimation, that such innovations could not be replicated on a grander scale, taking pressure off new sources of rock from places such as the Gravel Reach.

Recycled concrete was one way to significantly reduce overall concrete demand, Johnson said. Another was to consider rezoning changes that allowed houses to be built upward, rather than down. For example, Johnson said, there was no reason that houses could not be built on concrete footings or slabs. Doing so would mean no basements, but it would also mean no expensive concrete foundations. Another option, if basements are required, is to move to wood foundations in some structures. Built with two-by-six and two-by-eight lumber and faced with layers of treated plywood on the outside, such support structures could last long periods of time and be competitive with concrete, Johnson said, with the added benefit that they come from renewable materials.

But these latter innovations have their limits when weighed against certain inescapable realities. Wood-frame structures can only be built so high, whereas steel, concrete and glass structures are called skyscrapers for good reason. If high density is the objective, the latter always has inherent advantages over the former.

Nemkumar Banthia, a colleague of Johnson's at the University of British Columbia, had a similar focus in his work. He was particularly interested in extending the life of building materials.

"Concrete tends to be very brittle and cracks rather easily," he said.[39] For that reason, what it is reinforced with is very important. Typically, rebar or steel has been the reinforcement product of choice. But the knock against rebar is that it is easily corroded. As it corrodes, the entire structure weakens and eventually has to be replaced.

By the time of the Spring Bar controversy, Banthia said, design specifications typically required that reinforced concrete slabs retain their structural integrity for at least 75 years. Despite such requirements, Banthia said it is well within the realm of possibility that such structures could weaken far faster, perhaps in half that amount of time. In such cases, the culprit would be corrosion of the rebar hidden inside the slabs. All of this points to the need for more durable reinforcing materials made from polymers and other non-corrosive fibres, be they glass, graphite or carbon.

As the Fraser Valley and other regions urbanize, Banthia said, concrete and gravel demand will rise commensurately. Making concrete products that endure, then, is a critical element to any credible plan to stretch out the life of the province's finite gravel resources.

As Johnson's and Banthia's work demonstrates, the three Rs of responsible environmental stewardship — reduce, reuse and recycle — apply to more than pop cans and beer bottles. There are tangible ways to decrease how much gravel is needed and, in the process, reduce the pressure to mine sensitive environments and save vast amounts of energy.

It's energy conservation that has emerged as perhaps the most salient issue in the Fraser Valley and elsewhere during the first decades of the twenty-first century, because of the global risks to climate inherent in our increasing use of fossil fuels. But when combined with the downturn in the economy due to the global credit crisis, a unique opportunity has presented itself — to twin economic recovery with a broad greening of society.

The province of BC has already begun to lay the foundation for such a transition. In early 2008 it captured headlines across North America with its introduction of a carbon tax. The tax, while criticized in many circles for not being near enough to compel heavy fossil fuel energy users to change their ways, was nonetheless on the books. And its introduction, at a time of rapidly climbing fuel prices, forced a healthy debate about what policy interventions are needed to foster significant societal changes.

Tellingly, higher fuel costs and the new carbon tax prompted some industries to switch fuel sources. Some Fraser Valley greenhouse operations that once relied almost exclusively on natural gas to heat their massive glass structures moved to high-efficiency wood-fired systems. Gas prices subsequently came down, but the new wood-fired systems remained in place, and will likely act as a catalyst to the installation of more wood heat systems when gas prices track upward again. Much of the wood for the new systems came from the dead, beetle-killed pine trees in the upper Fraser's watersheds — trees that were slowly emitting greenhouse gases anyway, and that could be replaced by a new generation of carbon-storing trees. By switching energy sources, the greenhouse operators avoided paying a tax that will likely escalate with each passing year, and they helped to slow the depletion of non-replaceable fossil fuels.

Among the industries that initially got off "light" with the tax were cement makers, who emit massive amounts of greenhouse gases as they crush gravel to make their products. But cement makers, like all other fossil fuel users, must still pay the tax on all the fuel they purchase. A "cap and trade system" that sets caps on greenhouse gas emissions may also eventually force industries to reduce their emissions or purchase tradable "carbon credits" from low emitters. Already, in an emerging voluntary carbon credit market, some Fraser Valley greenhouse operators who have switched from natural gas to wood-fired heating systems have marketed such credits.

For gravel movers and users, the landscape is shifting, and their ability to innovate in how gravel is moved, used and re-used will be critical to their survival.

And key, too, to the survival of the Gravel Reach.

Salmon depend on the Fraser's sedimentary rock —until the end. LEIGH HILBERT PHOTO

X

Sinking Like a Stone: Gravel Mining and Reduced Flood Risk in the Heart of the Fraser

As the first decade of the new millennium drew to a close, continued gravel mining in the Heart of the Fraser rested on ever shakier justifications. Despite hundreds of thousands of tonnes of gravel that have been excavated from the river's gravel bars, the threat of flooding had been reduced not one iota. Yet the threat to some of the river's most important and endangered fish populations appeared to be on the rise, a rise that corresponded with the increase in gravel mining.

In March 2010, UBC's Michael Church issued yet another report to the provincial government on gravel removal in the reach of the river between Hope and Mission. Assessing the outcomes of large-scale gravel removal in the years following the 2004 launch of the five-year gravel extraction plan, Church saw much to be concerned about.

Removing gravel in the intense, round-the-clock "bar top scalping" operations that had predominated at Spring Bar and elsewhere in Sturgeon Reach since the program began had failed to influence how the river behaved and had "little effect on water levels" he said.[40]

The geographer was less equivocal on whether such operations had affected sturgeon and salmon saying "the actual sediment removals to date appear to have had no lasting impact on fish or invertebrate populations in the vicinity of the removals." But, he was not prepared to say that there had not been fish losses or habitat losses, noting that monitoring efforts during the years of gravel removal had been "insufficient to qualify this as a robust conclusion."

On this point, Church was not alone. More than a year earlier, provincial biologist Ross Neuman had sent a memo to provincial government environmental officials warning that good sturgeon habitat could be lost in the event that gravel mining proceeded at Little Big Bar, where tens of thousands of tonnes of gravel were slated for mining between January and March 2009. The actual mining did not occur until a year later, and when it did some 68,500 cubic metres of gravel or 5,480 dump truck loads of material were scooped up. But in the intervening year, not one provincial or federal regulator had heeded Neuman's warning that there was a correlation between declines in sturgeon stocks and the "initiation of large scale gravel removals in the lower Fraser."[41]

"Therefore," Neuman said, "it is prudent that further gravel extractions not proceed until the impacts to sturgeon and sturgeon habitat are better understood and until the risks of gravel extraction to the long term viability of white sturgeon in the lower Fraser has been determined."

In total during the five-year gravel mining effort, more than 1.14 million tonnes of gravel or 91,322 truckloads of material were removed in the name of flood control. And in 2010, in the first of what promised to be yearly extensions to the gravel-mining program, another 322,350 tonnes or 25,788 truckloads of gravel came off of Little Big Bar, Hamilton Bar and Gill Island combined.

The following year, however, brought at least a temporary reprieve. The Department of Fisheries and Oceans did not approve proposed gravel mining at Tranmer Bar near Chilliwack.

For the first time in more than a decade, no industrial-scale gravel mining took place in the Heart of the Fraser. This did not sit well with gravel industry advocates such as Chilliwack MLA John Les, who warned that the river would "plug up" and result in an "inevitable" flood.[42] Yet the arguments put forward by Church and others were beginning to hold sway. The gravel removals had not demonstrably reduced river levels.

While floods remain a risk, and will likely always be a feature of life along this stretch of the river, gravel mining appears to be singularly incapable of reducing their likelihood. To stand even a chance of being effective, gravel excavation would have to take place at a level far, far beyond what has occurred to date, with potentially disastrous results for the life that teems in and along the river.

As Church would note in an accompanying letter to his 2010 report: “We know from substantial experience that individual sediment removal short of the order of a million cubic metres will not substantially affect local water levels in the short term. But sediment removal on such a scale would very significantly disrupt the aquatic ecosystem. There is, furthermore, a concern that the current program pays too little attention to the potential ecological costs of sediment removal.”

With the temporary silencing of excavators and gravel trucks on Fraser’s gravel bars came a badly needed opportunity to rethink what is best for the region. If continued gravel mining isn’t the the key to flood control, what is the answer?

Of course, there isn’t one simple solution. Protecting the region from the threat of floods requires a multitude of responses, as does protecting the region’s unique ecology and providing a growing human population with all those rocks and stones that construction requires.

The region’s network of dykes and related flood-control infrastructure need improvement. It may be possible to engineer some river channels to improve stream flows at places like Harrison Knob. Zoning changes can promote urban density and reduce development in flood prone areas. Recycling requirements could lessen demand for new gravel. New building designs could reduce the need for so much gravel. Finally, the gravel we need can be found in places other than the Fraser itself.

All these solutions are in plain sight, if we look carefully at what’s around us, learn from what has gone before, and listen to the river.

NOTES

1. Aggregate Advisory Panel, *Managing Aggregate, Cornerstone of the Economy*, March, 2001.
2. Ministry of Energy, Mines and Petroleum Resources, *Construction Aggregates Resource and Stakeholders*, 2009 , http://www.empr.gov.bc.ca/Mining/MineralStatistics/MineralSectors/ConstructionAggregates/ReportsandPublications/Pages/ConstructionAggregatesResourceStakeholders.aspx.
3. Aggregate Advisory Panel, *Managing Aggregate, Cornerstone of the Economy.*
4. Levelton Engineering Ltd. *Lower Mainland Aggregates Demand Study. Volume 1: Aggregate Supply and Consumption. A report to BC Ministry of Employment and Investment*, June 19, 1996.
5. Levelton, *Lower Mainland Aggregates Demand Study. Volume* 1, 1996.
6. *Aggregates & Roadbuilding Magazine.* "1988 Top 20 Sand & Gravel Pits," 1988.
7. Kuhar, Mark, "Making the Most of the Coast." *Pit & Quarry Weekly Report*, May 1, 2004.
8. ABC Funds, "Polaris Minerals Corporation (TSX:PLS)," *Value Investigator*, April 13, 2007.
9. Borden, Charles. "Prehistory of the Lower Mainland." *Lower Fraser Valley: Evolution of a Cultural Landscape.* BC Geographical Series, Number 9, 1968.
10. Rosenau, Marvin, & Mark Angelo. *Sand and Gravel Management and Fish-Habitat Protection in British Columbia Salmon and Steelhead Streams.* Pacific Fisheries Resource Conservation Council, Background Paper, March 2003.
11. Borden, 1968.
12. Church, Michael, Darren Ham, & Hamish Weatherly. *Gravel Management in Fraser River. A report to the District of Chilliwack.* December 12, 2001.
13. Church, 2001.

14. Rosenau, 2003.
15. Church, Michael, & Hamish Weatherly. *Historical Changes of Minto Channel during the Twentieth Century. A report to the District of Chilliwack.* December 1, 1998.
16. Church, 1998.
17. BC Ministry of Environment. *Comprehensive Review of Fraser River at Hope Flood Hydrology and Flows – Scoping Study. Final Report.* October 2008.
18. Fraser Basin Council. *Flood Hazard Management.* 2009. http://www.fraserbasin.bc.ca/programs/flood.html.
19. *Ibid.* Flood boxes are used to channel water from smaller watercourses cut off by dykes to be returned to the main river. This is achieved through the installation of culverts that allow gravity fed water to flow downstream. Water from the main river cannot, however, flow back the other way because of flap gates at the ends of the culverts.
20. *Dredging News.* "Canada's Fraser river to be dredged." *Dredging News,* January 2, 2008.
21. Freeman, Robert. "Auditor asked to check gravel claims." *Chilliwack Progress,* June 11, 2008.
22. Kotyk, Mel. "Open Letter." Fisheries and Oceans Canada. January 25, 2007.
23. Church, Michael, & Darren Ham. *Atlas of the alluvial gravel-bed reach of Fraser River in the Lower Mainland showing channel changes in the period* 1928-1999. Department of Geography. University of British Columbia. 2004.
24. Hume, Mark. "Silencing the troubled waters: While the local chief says removing gravel poses no threat to salmon, opponents label the plan 'grotesque' and fear its impact." *Globe and Mail,* February 2, 2008.
25. Hume, 2008.
26. Millar, Robert. Personal communication. January 2009.
27. Church & Ham, 2004.
28. Church & Weatherly, 1998.
29. Fraser Valley Regional District. *Choices for our Future: Regional Growth Strategy for the Fraser Valley Regional District.* Fraser Valley Regional District, 2004.
30. BC Stats. *Quarterly Regional Statistics: Fraser Valley Regional District. Third Quarter* 2008. BC Stats, 2008.
31. BC Stats, 2008.
32. O'Neil, David (director, BC Stats, Population Statistics). Personal communication, 2009.
34. BC Stats, 2008.
35. Fraser Valley Regional District, 2004.
36. Fraser Valley Regional District, 2004.
37. Koch, P. "Wood versus nonwood Materials in US Residential Construction: Some Energy Related Global Implications," *Forest Products Journal* 42(5), 1992, pp. 31–42.
38. Johnson, Greg. Personal communication, January, 2009.
39. Banthia, Nemkumar. Personal communication, February, 2008.
40. Church, Michael. *Sediment management in Lower Fraser River: Criteria*

for a Sustainable Long-Term Plan for the Gravel-Bed Reach. A Report for Emergency Management BC, Flood Protection Program, Ministry of Public Safety and Solicitor General. March 30, 2010.

41. Neuman, Ross. E-mail memo to Environment Ministry personnel. January 2, 2009.
42. Freeman, Robert. "Critics question Fraser River gravel plan." *Chilliwack Progress*. March 7, 2011.

ADDITIONAL SOURCES

Ham, Darren, & Michael Church. *The sediment budget in the gravel-bed reach of Fraser River: 2003 Revision*. Department of Geography. University of British Columbia. October 20, 2003.

Rosenau, Marvin, & Mark Angelo. *Saving the Heart of the Fraser: Addressing Human Impacts to the Aquatic Ecosystem of the Fraser River, Hope to Mission, British Columbia*. Pacific Fisheries Resource Council, November, 2007.

Rosenau, Marvin. Letter to Auditor-General Sheila Fraser, October 2, 2008.

ACKNOWLEDGMENTS

The authors wish to thank Rudy North and Watershed Watch Salmon Society for their generous support and assistance in the research and writing of this book.

Thanks also to Derek Hayes for the use of the maps of the Lower Fraser, taken from his *British Columbia: A New Historical Atlas* (Douglas & McIntyre), that appear on page 4 and the back cover.

NEW STAR BOOKS LTD.
107 – 3477 Commercial Street, Vancouver, BC V5N 4E8 CANADA
1574 Gulf Road, No. 1517, Point Roberts, WA 98281 USA
www.NewStarBooks.com *info@NewStarBooks.com*

Cataloguing information for this title is available from Library and Archives Canada, http://www.collectionscanada.gc.ca/.

The publisher acknowledges the financial support of the Government of Canada through the Canada Council and the Department of Canadian Heritage Book Publishing Industry Development Program, and of the Province of British Columbia through the British Columbia Arts Council and the Book Publishing Tax Credit.

Cover by Rayola Creative
Cover photo by Charles Campbell
Printed on 100% post-consumer recycled paper
Printed & bound in Canada by Gauvin Press
First printing, October 2012

Watershed Watch Salmon Society's mission is to catalyze efforts to protect and restore BC's precious wild salmon. Through scientific expertise, strategic alliances, outreach programs, and innovative projects, Watershed Watch is at the forefront in sounding the alarm on threats to salmon and salmon habitat, and in prompting action to help them.

The authors have directed that the royalties from the sale of STURGEON REACH *be donated to Watershed Watch to further the society's aims.*

For more information about Watershed Watch, visit the society's website at www.watershed-watch.org/

FSC
www.fsc.org
MIX
Paper from
responsible sources
FSC® C100212